BEES

BEES
HEROES OF THE GARDEN

Tom Jackson

Published by Amber Books Ltd
United House
North Road
London
N7 9DP
United Kingdom
www.amberbooks.co.uk
Instagram: amberbooksltd
Facebook: amberbooks
Twitter: @amberbooks
Pinterest: amberbooksltd

Copyright © 2021 Amber Books Ltd.

All rights reserved. With the exception of quoting brief passages for the purpose of review, no part of this publication may be reproduced without prior written permission from the publisher. The information in this book is true and complete to the best of our knowledge. All recommendations are made without any guarantee on the part of the author or publisher, who also disclaim any liability incurred in connection with the use of this data or specific details.

ISBN: 978-1-83886-086-8

Project Editor: George Maudsley
Designer: Keren Harragan
Picture Research: Terry Forshaw and Justin Willsdon

Printed in China

Contents

Introduction	6
Social Bees	8
Solitary Bees	62
Bee Anatomy	100
In the Hive	128
Bees and Flowers	200
Picture Credits	224

Introduction

Never underestimate a bee. These unassuming insects go about their busy days, filling sunny flower beds and meadows with their tranquil, whirring buzz. Perhaps, upon seeing them, our first and only thought is, 'Oh good, it's not a wasp.' A bee is indeed less likely to wield its stinger, not least because honeybees are doomed to die should they use this ultimate weapon. What we forget to consider is that working largely

unseen, the bees are an incredibly productive group. Of course, honeybees are the first to spring to mind, workaholic insects that create complex societies for producing honey, one of the most valued of all foodstuffs. But, additionally, there is an impressive array of bees that live in other ways: mason bees build homes out of mud, while orchid bees collect exotic oils to woo suitors. And all this is powered only by foods made from pollen and nectar.

The world of bees will surely surprise and amaze, but we also need the hidden power of bees to pollinate and propagate our crops. We ignore threats to the world's bees at our peril.

ABOVE:
A Cape honeybee on a red aloe plant in southern Africa.

OPPOSITE:
Honeybees at a hive are a close-knit, hard-working team focused on producing honey for their young – and for human consumption.

Social Bees

Bees are famous for being sociable, living together in colonies or hives that contain hundreds or perhaps many thousands of individual bees. All the bees are working together for the common good under the rule of a single, all-powerful queen. The most familiar social bees are honeybees, but bumblebees and stingless bees also live this communal life. Nevertheless, the social bee species constitute a small fraction – less than 1,000 species out of more than 20,000 – of the broader bee group, Anthophila. The bee's gregarious way of life, which is shared with their more beastly cousins, such as yellow-jacket wasps and ants, is the most extreme form of social group seen in the animal kingdom. The queen is the only reproductive female, and the rest of the colony are her infertile daughters. These females work to raise their sisters to swell the ranks of workers, and on occasion a few brothers, too. The system works well because of a genetic quirk in the way a bee's sex is determined. It is normal for offspring and siblings to share half their genes, and this level of relatedness is enough for them to be altruistic toward each other, up to a point. However, male social bees have half the number of genes as females, and this skews genetic relatedness. Worker bees share three-quarters of their genes with their sisters, which includes the next generation of queens. This stronger genetic link is enough for a female to abandon her own breeding opportunity and work to boost that of her sisters.

OPPOSITE:
Spot the queen
The queen – shown here in a domestic hive in Italy by the spot of green paint – is considerably larger than her workers. The size difference is due to enlarged ovaries for producing fertile eggs.

OVERLEAF:
Carniolan honeybee
Named after the coastal region of Slovenia, the Carniolan honeybee is the main kind of honeybee in the Balkans and northern Italy, where there is a tradition of giving hives a brightly painted pattern. This subspecies of the western honeybee, nicknamed 'carnies', is known worldwide for producing healthy hives, resistant to pests.

ABOVE:
Honeycomb
All honeybees build a nest comprised of hexagonal cells. This is the near optimal compromise between the nest having high rigidity as well as maximizing the volume.

RIGHT:
Cape honeybee
The Cape bee is a subspecies of *Apis mellifera*, the western honey bee. It clings to the very tip of southern Africa around the Western Cape region, where it is a crucial pollinator in local agriculture and natural habitats. This forager is gathering pollen from an evergreen shrub called orange jasmine.

Sneaky breeder
A Cape honeybee collects nectar from a Cape aloe. Cape honeybees are unique among the western honeybees in that the female workers – or at least a few of them – can lay fertile diploid eggs. Diploid eggs have a double set of genes and so develop into females – effectively the cloned daughters of the 'laying worker'. In areas where the Cape honeybee lives alongside other subspecies, the laying workers can sneak eggs into neighbouring nests where they are raised by the unsuspecting – and unrelated – worker bees.

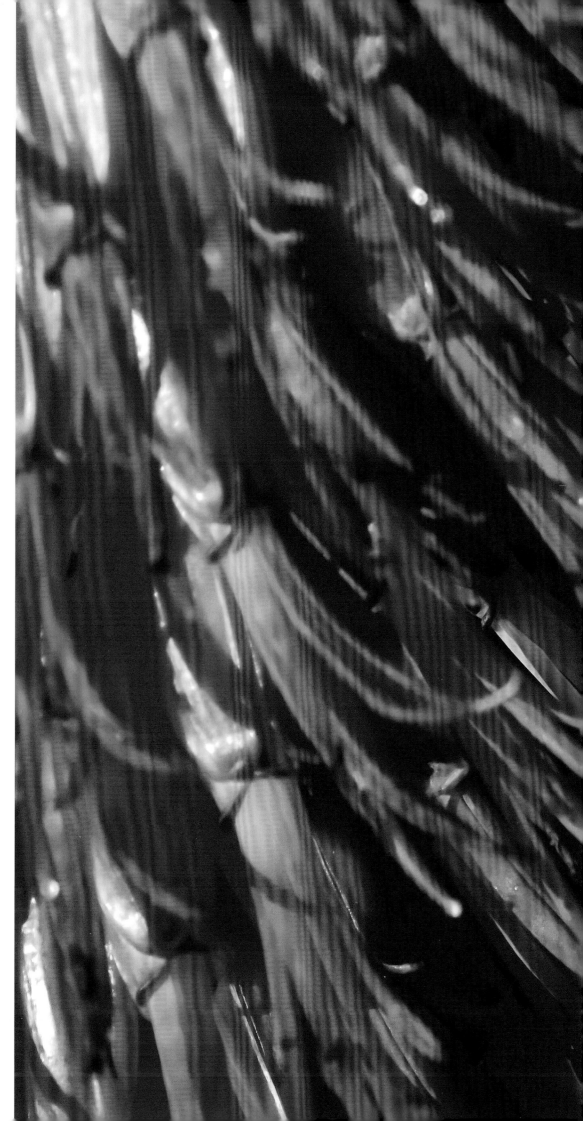

ABOVE:
Caucasian honeybee
Named for the Caucasus Mountains, this subspecies, *Apis mellifera caucasia*, is also common in Turkey and the northern parts of the Middle East. They do not do well in colder climates, and are not well suited to beekeeping because they are most active at the end of summer, whereas most nectar sources become abundant in spring.

OPPOSITE:
Cretan honeybee
A worker bee of the subspecies endemic to Crete is gathering nectar to take back to the hive. This subspecies, *Apis mellifera adami*, was only identified in 1975, and has a varied genetic profile. This suggests it is actually a hybridization of Mediterranean strains that has developed on the Greek island due to beekeepers bringing in new varieties of bee.

Giant honeybee
This species, *Apis dorsata*, lives across southeast Asia from the foothills of the Himalayas to the Wallace Line, which divides Asian wildlife from Australasian fauna. At about 17–20mm (0.7–0.8in) long, the giant name is apt, it being 70 percent bigger than most other honey bees.

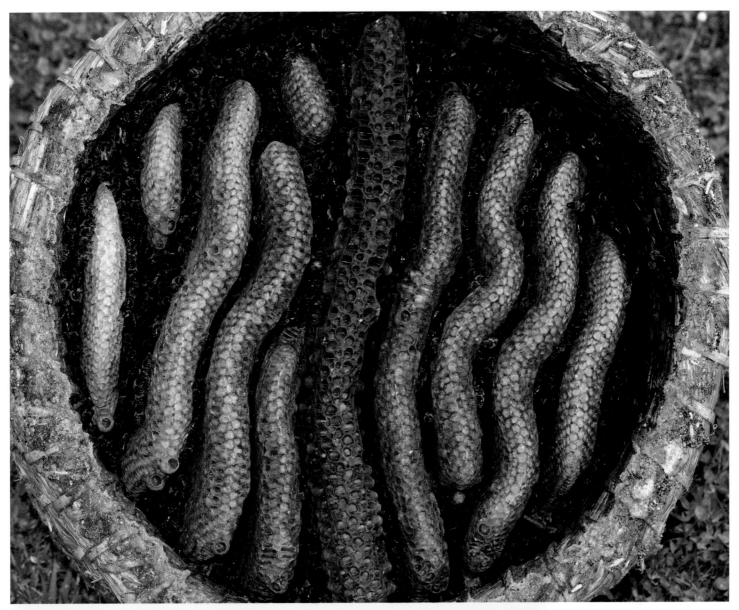

OPPOSITE:
A queen and her court
A western honeybee queen is surrounded by her court, laying eggs into the 'broodcomb' – the part of the wider honeycomb set aside for raising young.

ABOVE:
Natural and artificial
Here, a honeybee colony has built an amorphous honeycomb in a symmetrical artificial hive in the Netherlands.

LEFT:
Drinking up
As their name suggests, honeybees eat mostly honey, and exclusively for those bees confined to the hive. However, these foraging workers are pausing mid-flight to lap up a drink of water with their feathery, tongue-like mouthparts.

RIGHT:
Eastern honeybee
The species *Apis cerana* is native to South and East Asia. Only the Indian subspecies has been domesticated for honey production. The eastern species is very similar to the western honeybee, *A. mellifera*, in size but has discernibly thinner yellow stripes.

OVERLEAF:
Black dwarf honeybee
A close-up of the head and mouthparts of the black dwarf honeybee, *Apis andreniformis*, shows the feathery tongue-like mouthpart called the glossum. This is used to slurp up liquids – nectar, honeydew, water, and, of course, honey. *A. andreniformis* lives in the forests of Southeast Asia. It is about 6.5mm (0.25in) long, about half the length of a domestic honeybee.

LEFT:
Swarm
A giant honeybee swarm hangs from a tree branch. This wild swarm will have recently split from a parent colony that was growing too large. A new queen is protected at the centre.

ABOVE:
Domestic animal
Workers busy themselves at the entrance to a domestic hive in Italy. Honeybees have been domesticated for at least 4,500 years, probably first in ancient Egypt. Long before that, our ancient ancestors collected honey from wild nests.

OVERLEAF LEFT:
African honeybee
A swarm of *Apis mellifera scutellata* gathers on a pipe. This is the main subspecies in the lowland areas of southern and central Africa. It is likely that the subspecies is an ancestral form of the whole western honeybee species, which spread out from East Africa about 7 or 8 million years ago.

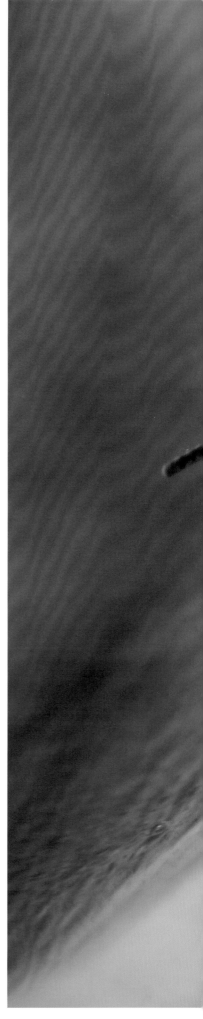

Dwarf bee
A red dwarf honeybee, *Apis florea*, laps a drop of water on the lip of a sink. This species broadly overlaps with that of its similarly diminutive close relative, the black dwarf honeybee, in Southeast Asia. However, the red species is also found in much of the Indian subcontinent and even parts of the Middle East that have rainfall high enough for forests to grow.

Plant foods
As with other social bees, the bumblebees consume nectar and pollen, with foragers carrying it back to the nest to share with the workers and to feed to brooding larvae. However, unlike honeybees and stingless bees, the insects do not produce honey in the true sense. They provision their young with small amounts of regurgitated nectar and chewed pollen, which forms a sweet paste.

What's the buzz?
The name bumblebee is said to have emerged from the low, persistent buzz these insects make in flight, plus their oddly slow and ponderous demeanour, as if they are bumbling from flower to flower. At least, that is one of a few theories behind their distinctive name.

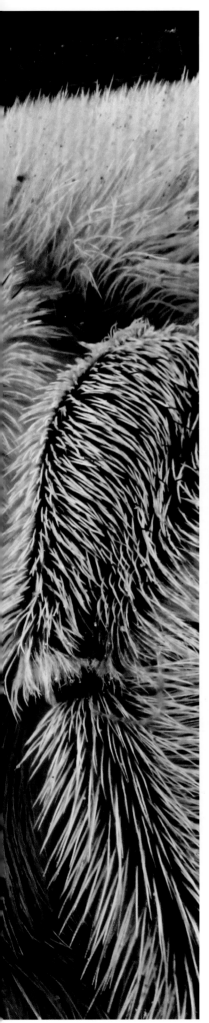

LEFT:
Eyes not ears
Bumblebees cannot hear – they have no ears – but can probably detect vibrations through other body parts. The eyes and antennae, which detect chemicals, are the primary sense organs.

ABOVE TOP:
Flight system
Bumblebees beat their wings 200 times a second. Each beat is not controlled by a single nerve signal – that would be too slow a command system. Instead the flight muscles vibrate like a taut string and this creates the high frequency of beats.

ABOVE BOTTOM:
Chunky
Bumblebees have a thick-set appearance. They are about 50 percent larger than honeybees (although one Chilean species is 4cm, or 1.6in, long and called the 'flying mouse'), but they appear larger than that thanks to the thick coat of setae, or short bristles, that covers the body.

LEFT:
Warming up
Bumblebees appear earlier in the spring than honeybees and thrive in colder locales. This is due in part to the insulation afforded by their thick 'fur' coats. However, the cold-blooded insects also warm themselves up by absorbing the sun's heat with their dark bodies and by shivering the flight muscles to generate some internal warmth.

OVERLEAF:
Pollinators
As with all social bees and many other flying insects, bumblebees are important pollinators, ensuring that plants are able to transfer pollen needed to breed and develop seeds. Flowers produce nectar to attract the bees and, in return, the bees transport sticky pollen grains on their bristled bodies – unwittingly – from flower to flower.

Old and new worlds
The tree bumblebee, or *Bombus hypnorum*, lives in Europe and northern Asia, spreading all the way to the Pacific coast of Siberia. Bumblebee species also live in North and South America, plus the northern fringe of Africa. However, they are not found south of the Sahara or in Australia.

Stingless bees
Most social bees are so-called stingless bees. These insects actually do have stingers but they are so small as to be useless. The stingless bees, of which there are about 500 species, often have a more waspish look, with a narrow 'waist' between thorax and abdomen.

Brood parasite
Like its avian namesake, the lemon cuckoo bumblebee, *Bombus citrinus*, from the northeast of North America, does not raise its own young. Instead it sneaks eggs into the brood chambers of other species of bumblebee and fools their workers into doing it for them. This mode of reproduction is called brood parasitism.

RIGHT:
Bite not sting
A squad of older members from a stingless bee colony in Brazil (the species is *Scaptotrigona xanthotricha*) are seconded as soldiers and rush to the broad entrance of the nest. Without a sting the bees will bite, and some species have a nasty acidic saliva that aggravates the wound.

OVERLEAF TOP LEFT:
Nests
Stingless bees make their nests in cavities in rotting wood, dead trees and underground. The nests generally have a single entrance – and exit – point that is built to allow a large number of the colony to leave at the same time.

OVERLEAF MIDDLE LEFT:
Australian honeybee nest
There are 14 species of stingless bee in Australia. Indigenous Australians practised beekeeping with stingless bees for thousands of years, and also collected their honey from wild nests. Stingless bee honey is thinner and more runny than that of true honeybees, and its unusual mixture of sugars is proven to be actively healthy compared to other sources of sweetener.

OVERLEAF BOTTOM LEFT:
Honey pot
The stingless bees make honey from a mixture of nectar and pollen – as do all honeybees. The stingless variety is stored in the nest in little spherical honey pots, seen here being pumped out. The bee makes the honey by ripening a droplet of regurgitated nectar and rolling it in the mouthparts to dehydrate it into a syrup.

OVERLEAF RIGHT:
Resinous
A *Trigona* stingless bee from Brazil busies herself with nest maintenance. She is using a paste made from resins collected from nearby plants to construct the nest.

Beekeeping
Stingless bees have been semi-domesticated in Australia and South and Central America, most notably by the ancient Maya of Mexico and Guatemala. An established colony was collected from the wild by these beekeepers. They looked for nests in logs that could be carried and kept near to their homes.

RIGHT:
Fake signal
This Asian species of stingless bee is mimicking the warning colours of wasps and honeybees to fool predators into thinking it carries a stinger.

FAR RIGHT:
Stingless bees
Stingless bees are most abundant in South America, with three-quarters of species living there. However, they are also represented in Africa, Asia and Australia. The honey from this Asian species, *Heterotrigona itama*, is being trialled as a possible anti-obesity food supplement.

OVERLEAF:
Bearding
A gang of honeybees hang out away from the colony. This activity is known as bearding – mostly with a larger group than this that looks like a beard (and is sometimes worn as one by showmen and daredevils). The behaviour occurs before a swarm or also when the internal temperature of the hive is too high.

Solitary Bees

Despite what you may have heard, it is not normal for a bee to live in a large group. Most species – many thousands of types found worldwide in the Anthophila group – are solitary creatures that keep themselves to themselves. As with honeybees and bumblebees, these insects get busy collecting pollen and nectar, but each female builds her own nest and prepares it for her own eggs. How they do this varies from group to group and, as we'll see, the bees use a range of ingenious techniques. Generally this hard work is done during the spring and summer. This sets the stage for eggs to hatch during the autumn, and the larvae – the maggot-like young forms – spend the winter eating their way through a food cache. This is a big ball of pollen and regurgitated nectar that dwarfs the newly-hatched young one. As the weather warms, the now chubby larva will pupate and is soon ready to emerge as an adult.

The males generally emerge first and are just as busy as the females, but less productive. They tend to be smaller than the females due to the shorter development time, but they also need to build a nest or work to provide food for the young. Instead they are simply on the lookout for a mate, which they do by hovering near a nest-building site, or displaying on a perch, advertising their status using a distinctive set of odours.

Solitary bees are not always alone. Some species live in communal settings, with females helping each other out, a hint perhaps of how social bees first appeared.

OPPOSITE:
Orchid bees
Metallic-green orchid bees are so named because they are the sole pollinators of some orchid flowers, although they do visit other kinds of blooms to harvest a nectar with a long proboscis.

PREVIOUS PAGES:
Scent collectors
Male orchid bees have comb-like brushes on their legs, which they use to trap the unique fragrances of orchids, other flowers and even rotting wood. This behaviour gives the male bee a distinctive odour, which attracts females during the breeding season.

OPPOSITE:
Carpenter bee
A Caucasian carpenter bee, or *Xylocopa valga*, crawls from its nest in a dead log. These solitary bees are named for the way they cut a nest into wood. There are about 400 species found worldwide. The generic name *Xylocopa* means 'wood cutter'.

ABOVE TOP:
Big and shiny
A violet carpenter bee, *Xylocopa violacea*, seen across Europe and Asia, collects pollen and nectar from a flower. The bee's size – about 25mm (1in) – and vociferous buzz means the insect is superficially mistaken for a bumblebee, but can be told apart by their metallic-looking bodies.

ABOVE BOTTOM:
Valley carpenter bee
This American species, *Xylocopa sonorina*, is found from west Texas to the Carolinas. It is also known as the teddy bear bee, because although the female sports a black metallic look, the males are clothed in a golden brown coat of thick hairs and have compelling green-blue eyes.

LEFT:
Southern carpenter bee
This view from the top of the head of a male *Xylocopa micans* from the American Deep South and Central America shows the bee's ocelli, or simple eyes, which detect light and dark above – perhaps the shadow of a predator, such as a woodpecker. We know it is a male because of its yellow fluffy hairs. His large compound eyes – bigger than a female's—are used to scan for potential mates.

OVERLEAF LEFT:
Home sweet home
Carpenter bees spend the winter – mostly in the larval stage – in a nest dug into wood, such as the branch of this acacia tree. Both males and females spend most of the summer outside searching for mates and collecting food for brooding young.

OVERLEAF TOP RIGHT:
Scraping and gouging
Carpenter bees will recycle old nests wherever possible, but if none are available, a female will dig a new one, using her mandibles to cut through the wood. Males do not dig nests.

OVERLEAF BOTTOM RIGHT:
Division of labour
Carpenter bees lay their eggs in side chambers dug at right angles from the main nest. Each chamber is provisioned with a ball of pollen and regurgitated nectar. Often groups of females will nest close together and divide up the duties of finding food and digging nests.

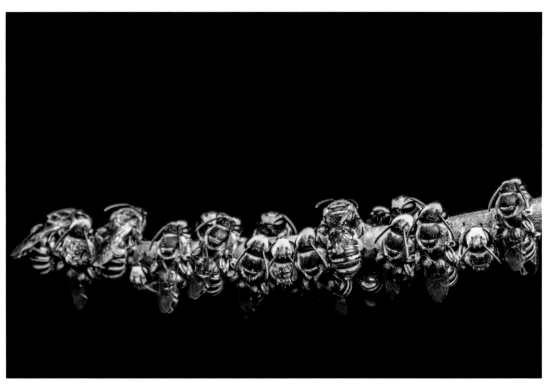

OPPOSITE TOP:
Cuckoo bee
These waspish-looking bees do not raise their own young. Instead they are kleptoparasites like their avian namesakes, sneaking into nests and laying an egg in brood cells alongside the host's. Once hatched, the larva cuckoo bee kills their nursery mate – and may eat it – and is thus raised by the stolen labours of an unsuspecting host.

OPPOSITE BELOW:
Resting
A male cuckoo bee clings to the anther of a flower with its legs tucked up and holding on with its pincer-like mandibles. This is called sleeping position, although the sleep in bees is nothing like we experience.

ABOVE:
Neon cuckoo bee
This species, *Thyreus nitidulus*, is from Australia. As a general rule, cuckoo bees parasitize species that look similar to them so they can pass undetected among them.

Big mouth
This is a member of the *Nomada* genus, the largest group of cuckoo bees, and with 850 species one of the largest genus of any bee type. As this posture shows, the mandibles of cuckoo bees are especially strong. The larval form has fearsome mouthparts that are the weapons used to kill host larvae in the first moments of life. After developing into an adult, the mandibles are more petite (but not lacking in power) and are devoted to eating pollen and nectar.

OPPOSITE:
Digger bee
As the name suggests, these bees build burrows in the dry earth as shelters and to construct waterproof brood pots for their young. This female *Melissodes* species is collecting pollen from a prickly pear as food for herself and for provisioning her eggs.

LEFT TOP:
Digger bee
Most of the digger bees belong to the Anthophorini tribe, which is a worldwide group containing about 750 species. However, the digging behaviour is also seen in more distantly related groups, such as the Centridini, which are distinct for collecting oils from flowers instead of pollen.

LEFT BOTTOM:
Longhorn species
Members of the *Melissodes* genus are known as long-horn digger bees. However, only the males live up to this name, with antennae that are twice as long as a female's.

OVERLEAF:
Urbane digger
The urbane digger bee, *Anthophora urbana*, lives in the United States and Mexico, and is especially common in alpine meadows. This is a male bee as indicated by his white 'bearded' face.

Leafcutter
A female leafcutter bee hauls a sheet of leaf cut from a nearby plant back to her nesting site. She has chosen a hollow stick in which to build the nest and will roll this and subsequent leaf fragments into a compartmentalized tube, subdivided into brood chambers for her larvae.

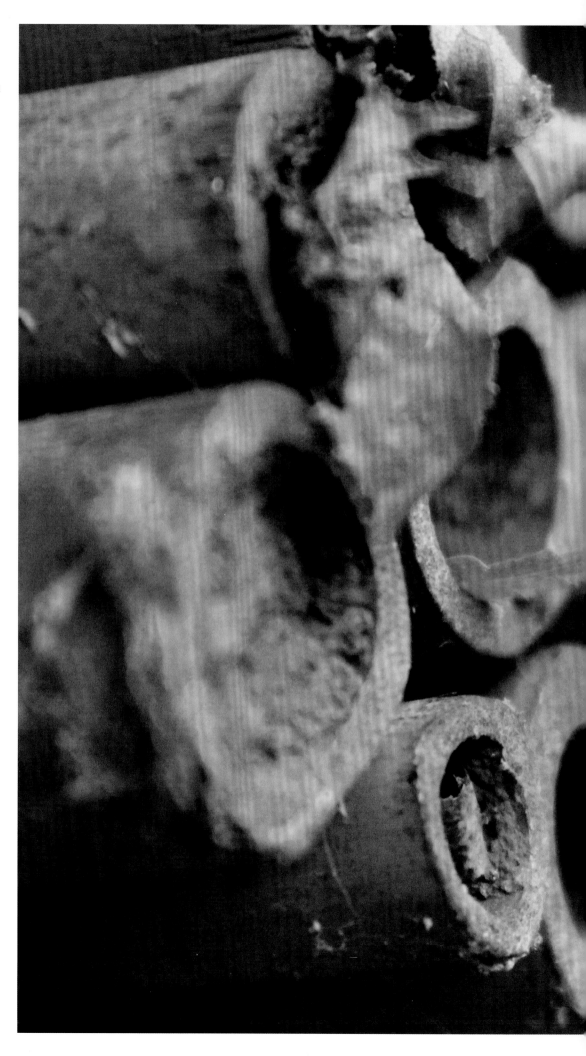

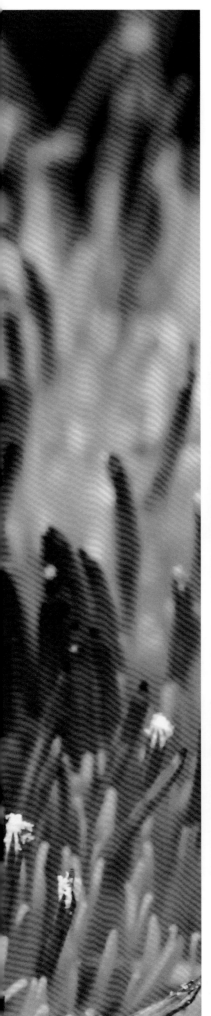

LEFT:
Pugnacious leafcutter bee
This North American species, *Megachile pugnata*, is a common visitor to sunflowers and related plants. It is collecting pollen and nectar from a thistle flower. It will use this food to provision brood chambers back at the nest.

BELOW TOP:
Leafcutter bees
A leafcutter bee sleeps on a grass flower. There are around 1,500 species of leafcutter bee, mostly in the *Megachile* genus, and found worldwide. This group includes *M. pluto*, Wallace's giant bee, and is the largest bee of any kind. The females of this Indonesian monster grow to nearly 4cm (1.6in) long.

BELOW BOTTOM:
Chew and slice
Leafcutter bees use their mandibles to slice their way through leaves and sometimes petals to create a very roughly rectangular piece, a bit longer than it is wide, that can be rolled up back at the nest site.

RIGHT TOP:
Mud builders
Mason bee larvae are seen feasting on pollen balls gathered by their mothers before they laid their eggs. This was done on top of the food, when both were sealed away in their private mud chambers. The larvae spend the winter eating the blob of pollen and nectar, before pupating and emerging as an adult in spring. The males will leave first and wait near the nests, which are often clustered in suitably muddy locations, for the female to leave. Each one is a potential mate.

RIGHT MIDDLE:
Red mason
The *Osmia rufa* species, commonly know as the red mason bee thanks to its gingery appearance, lives across temperate Europe from England to the fringes of western Asia.

RIGHT BOTTOM:
Blue orchard bee
A pair of blue orchard bees, *Osmia lignaria*, prepare to land on a pollen-rich flower. The females build a nest in natural crevices or hollows, near to a supply of soft mud, which is used to line the nest.

OPPOSITE:
Mason bee
As with all bees, mason bees are crucial pollinators, spreading pollen from bloom to bloom. Some North American species, including the orchard bees, are bred specifically to help pollinate commercial crops.

OVERLEAF:
Northerners
Often a mason bee will occupy the nest hole made by a carpenter bee. There are about 350 species of mason bee living in the Americas and across Eurasia. Most of them are found in the Northern Hemisphere. They are close relatives of the leafcutter bees and carder bees, which chew up leaves into a pulp for living nests, and resin bees, which collect the sticky secretions of trees for that purpose.

LEFT:
In a hole
Mining bees dig communal nests into soft soils, where several females will share the work. The spoil of loose soil creates a chimney around the entrance. There are 1,300 species of mining bees, mostly in the *Andrena* genus, which are found worldwide, except South America and Australasia.

BELOW:
Smaller sex
Mining bees are small, ranging from 8–15mm (0.3–0.6in) in length. Male mining bees are smaller and more slender than the females. The males, such as this one catching a bite to eat on a zinnia flower, are not required to do any physical labour building and provisioning of the nests. Mining bees can be distinguished from other kinds by the short velvety tuft of hairs between the eyes.

Summer exit
After spending winter in their sealed-off, subterranean brood cells, the newly pupated adults dig themselves out once the temperature above ground reaches about 20°C (68°F).

ABOVE TOP:
Plasterer bee
A *Hylaeus* species visits a field scabious plant. This is one the genera in the Colletidae family, which totals more than 2,000 species of plasterer bee. The insects earn this name from the way they finish their underground nests with a secretion from their mouths, which creates a smooth internal lining. The lining is a natural form of polyester.

ABOVE BOTTOM:
Spring mining bee
The *Colletes cunicularius*, the spring mining or vernal bee, is one of the most widespread of bee species. It is found in sandy habitats from the British Isles to the Pacific coast of Siberia. Members of the Colletidae family are found worldwide but are most abundant in South America and Australia.

RIGHT:
Late arrival
This large Colletes species – large being over 10mm (0.4in) – emerges from its underground nest in late summer. They are active until the end of autumn, by which time they will have provisioned their eggs in nests built in loose soils.

Sweat bees
Members of the *Halictidae* family are known as sweat bees. This is because some of them, especially the smaller species, are attracted to sweat. This species is more active at dusk or dawn, using their large ocelli, seen here as dark, shiny spots on the head. The ocelli are simple eyes that help the bees orientate themselves in lower light levels.

RIGHT:
Pollen fan
Pollen and nectar are the only source of food for the sweat bees, or *Halictidae*, and the females collect it to provide a food store for larvae. The eggs are laid in underground nests.

BELOW:
Metallic green
Most sweat bees have a shiny green body. However, many species also show off the warning stripes of other kinds of bee. This is an example of mimicry, where all striped bees and wasps benefit from sending out the same signal that warns attackers to beware of a sting. Sweat bees can sting but the weapon is not a powerful one.

OPPOSITE:
Big family
The sweat bee family comprises more than 3,100 species, making it the second largest bee family. Most species are small for bees, measuring between 6 and 8mm (0.24 and 0.31in).

Social animals
Most sweat bee species live solitary lives, although they may gather in large groups when the conditions allow. However, some species exhibit eusocial behaviours, similar to those of honeybees, with a queen and workers. Some of these species can only live in this way; others adopt this social system when the supply of food and water make it the best option.

Bee Anatomy

Bees have a highly distinctive body plan, with a set of sleek wings, a lithe but sturdy body, often highlighted with thick pale stripes, a narrow 'waist' mid-body, and a big, busy head. The bee shares these characteristics with a wider group of related insects called the Hymenoptera. This order also contains ants, wasps and sawflies, which outnumber bees ten to one.

The name Hymenoptera could mean 'membrane wing', accounting for the near-transparent wings. However, a better translation is 'married wings', evoking Hymen, the Greek god of marriage. All members of the group have four wings, one large pair in front of a smaller set behind. Each hindwing is attached by hooks to its forewing partner, creating a structure that works as a single flight surface. Additionally, the bees share their tight 'waist' feature with their cousins. In fact, this is much more pronounced in ants and wasps.

Bee anatomy, especially such fine details as the segmentation of the face and abdomen, shows that the bees belong to a single distinct branch of the Apocrita suborder. This group includes ants and wasps and makes up the great majority of the Hymenopterans. Only the sawflies are left out. It seems logical that bees are a closer relative to their more devilish cousins, the yellowjacket wasps – after all, they are of similar stature, both sting and live in colonies of comparable size. However, genetic evidence points to bees being more related to ants.

OPPOSITE:
In sections
As with all insects, a bee's body is organized into three main sections: the head, thorax and abdomen. The head hold the mouthparts and sense organs. The wings and legs are on the thorax, while the abdomen holds most of the internal organs.

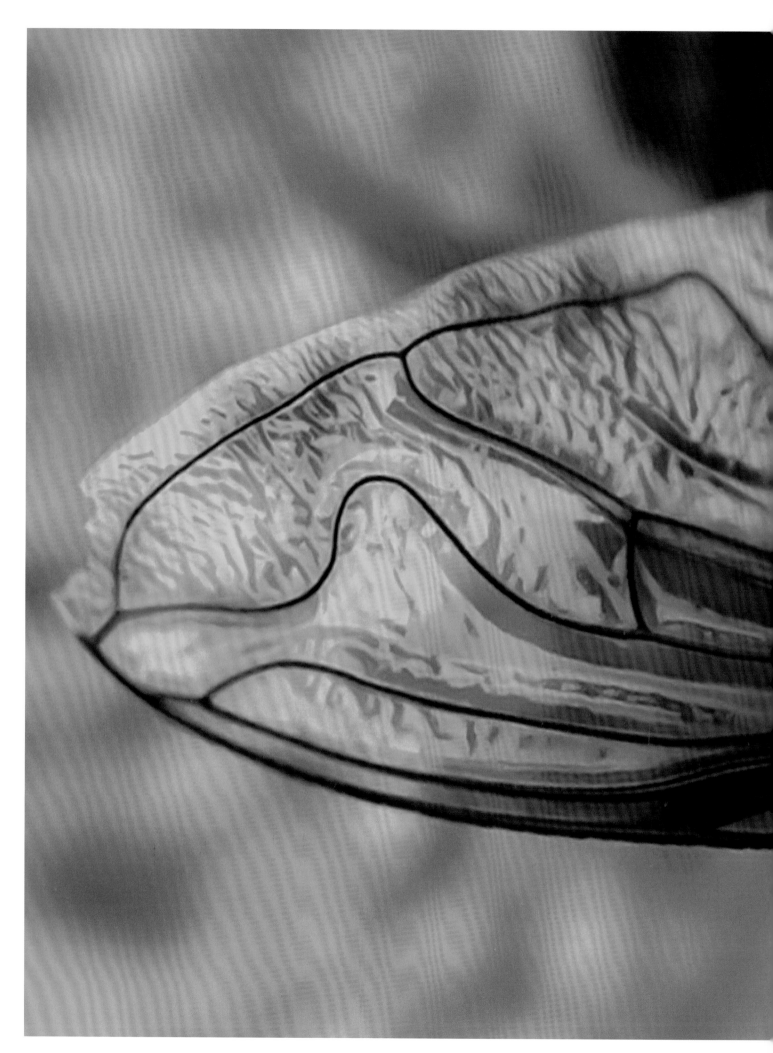

Distinctive vein
Bee wings have a distinctively sturdy vein along the leading edge, seen here at the bottom. This part cuts through the air as the wing flaps up to 200 times per second. It also has a graceful curved shape, while other hymenopterans have more elongated wings.

RIGHT:
Setae
Bees are often described as hairy when of course only mammals like us grow true hairs. The proper term for an insect hair is a seta, or setae in plural. These are projections of chitin, the hard but flexible material from which the insect's exoskeleton is made.

BELOW:
Legs on show
Bees have six legs, as do all insects. All three pairs are attached to the midbody thorax section, and they are made up of several jointed sections. The muscles that move the legs – and other body parts – are anchored to the inside of the exoskeleton.

LEFT:
Mouthparts
Bees have a surprisingly uniform set of mouthparts. The most significant are those that protrude into a flower to slurp up nectar and handle pollen. They include an outer pairing of maxillae, and inner set of two labial palps, and a central tongue-like glossa that is covered in feathery bristles. Above this set of feeding organs are the pincer-like mandibles, used for chewing on pollen.

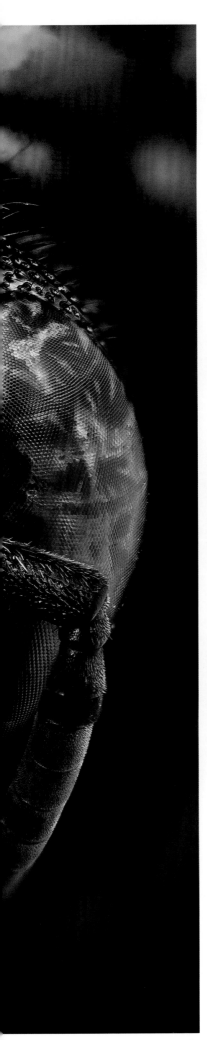

LEFT:
Elbows on the head
As this head-on view of a large, parasitic orchid bee, *Exaerte smaragdina*, shows, the antennae (or feelers) of all bees are elbowed, or have a joint half way along.

ABOVE TOP:
Cocoons
Bees undergo a complete metamorphosis from worm-like larva to winged adult. Just like the caterpillar hides away in its chrysalis before emerging as a butterfly, these honeybee workers are pupating immobile in cocoons, putting the finishing touches to their adult forms, before cutting themselves out.

ABOVE BOTTOM:
Only eyes for one thing
This hatching western honeybee (*Apis mellifera*) is a drone. We can tell from its enormous eyes, which it uses to scour the skies for a young queen to impregnate. Famously, honeybee drones make one nuptial flight and they mate with such enthusiasm that they literally explode as they ejaculate.

Segmented feelers
Bee antennae are segmented beyond the elbow. A male bee has 13 segments, while the female always has 12.

LEFT:
Vision
Bees do not see the world in the same way as us. Their large compound eyes are made up of thousands of individual tube-like lenses called ommatidia. These each send an image signal to the animal's brain, which is used to build up a single picture.

ABOVE TOP:
Four wings
This action shot shows the bee's four wings frozen mid-beat. Bristle-like hooklets on the front edge of the hind wing toward the top, known as hamuli, attach it to the underside of the forewing.

ABOVE BOTTOM:
Structural colour
The iridescent colouring of many bees' wings and bodies is not due to chemical pigments. Instead, the rainbow, oil-pattern effect is produced by the way some light reflects off the surface while other beams penetrate through and reflect from deeper down. These two light sources then interfere to make the shimmering patterns.

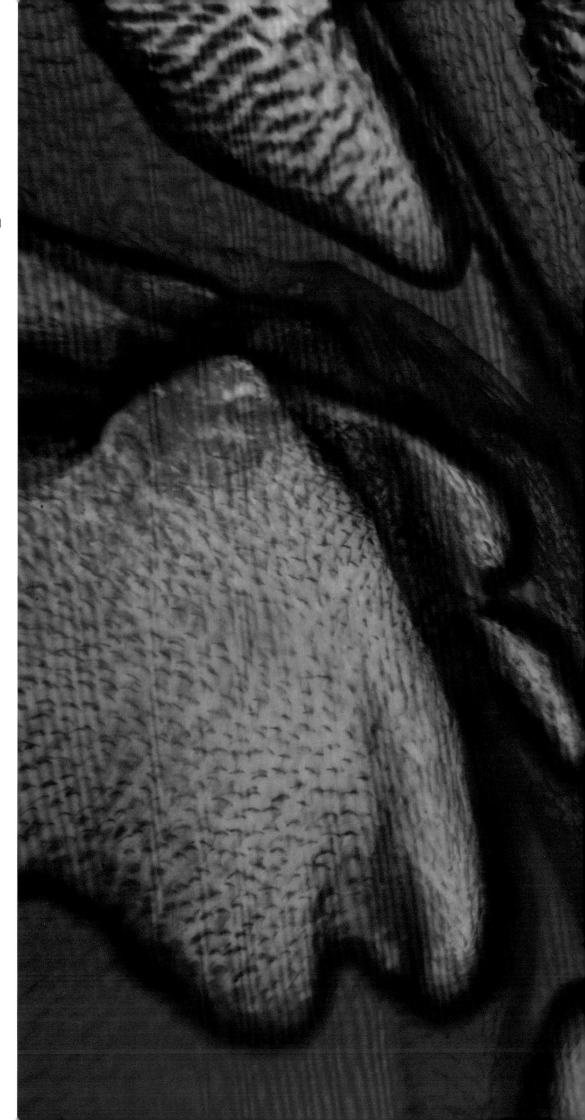

Wing structure
A bee wing under a microscope shows that it is made from an upper and lower surface of chitin – the same stuff as in the rest of the exoskeleton. The wing is then stiffened by an inner layer of haemolymph, which is the insect equivalent of blood, which carries oxygen and nutrients around the body. Instead of pumping through blood vessels, this colourless liquid washes through the body cavity.

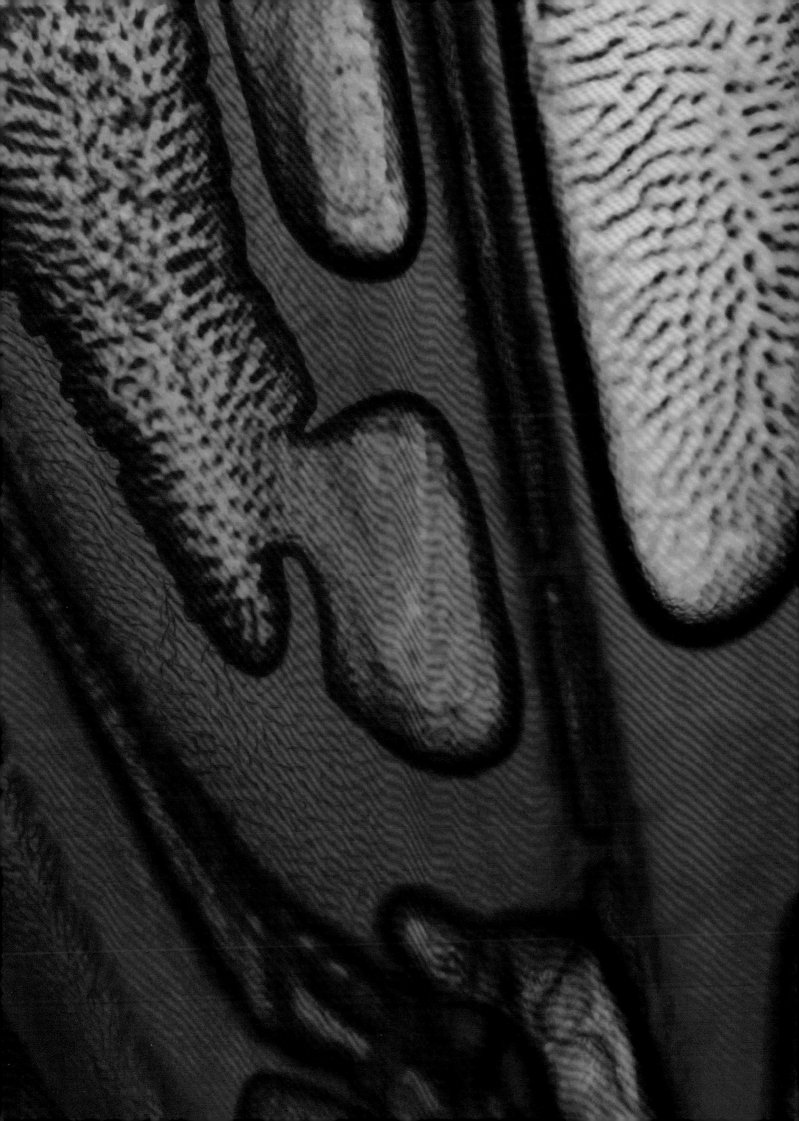

ABOVE:
Male sense
Despite being misnamed as a feeler, the antenna is an array of sophisticated sensory systems that detects chemicals and picks up vibrations. This male long-horned bee has longer antennae than the females because it is maximizing its ability to pick up the odours of potential mates.

OPPOSITE TOP:
Hearing
The bee's antenna has fine hairs that are linked to nerve cells. As these hairs waft in the currents of air that pass by, they are able to send motion information to the brain and also something like hearing signals, as loud noises create pressure waves in the air. Bees do not have true ears.

OPPOSITE BOTTOM:
Touch
Bee antennae also work as feelers, with nerve-linked mechanoreceptors clustered near their tips responding to touch and pressure.

LEFT:
Motion sensors
The compound eye does not provide a detailed view of the surroundings. However, it is very good at picking up motion as an object moves across the eye.

ABOVE:
Vein support
The wing veins are a deeper and thicker region of the upper and lower cuticle that runs through the wing. This creates a rigid structure supporting the shape of the wing, with cross-branches adding more strength.

Clawed feet
A honeybee effortlessly grips to a smooth surface using its small but strong tarsal claws. There are three on each foot.

BELOW TOP:
Cleaner
One unique feature of bee morphology is a brush-like structure on the front foreleg, which is used to clean the antennae and mouthparts.

BELOW BOTTOM:
Modified egg tube
All bee species have stingers – even stingless ones, although theirs are just tiny. Only a female bee can sting, because the stinger is part of the ovipositor, or egg-laying system.

RIGHT:
Grooved face
The bee's face is made up of two plates. The clypeus is like a lid sitting over the mouthparts, while the frons above it is more like a forehead. There is a distinct articulated groove.

Abdomen

There are few body parts as distinctive as a honeybee's abdomen. The yellow bands on a black background are an example of aposematism, where the animal is sending out a signal that attackers may come to harm. It probably evolved through the process of Mullerian mimicry where many species adopted the same signal and thus all species benefit from the other unrelated species spreading the message widely on their behalf.

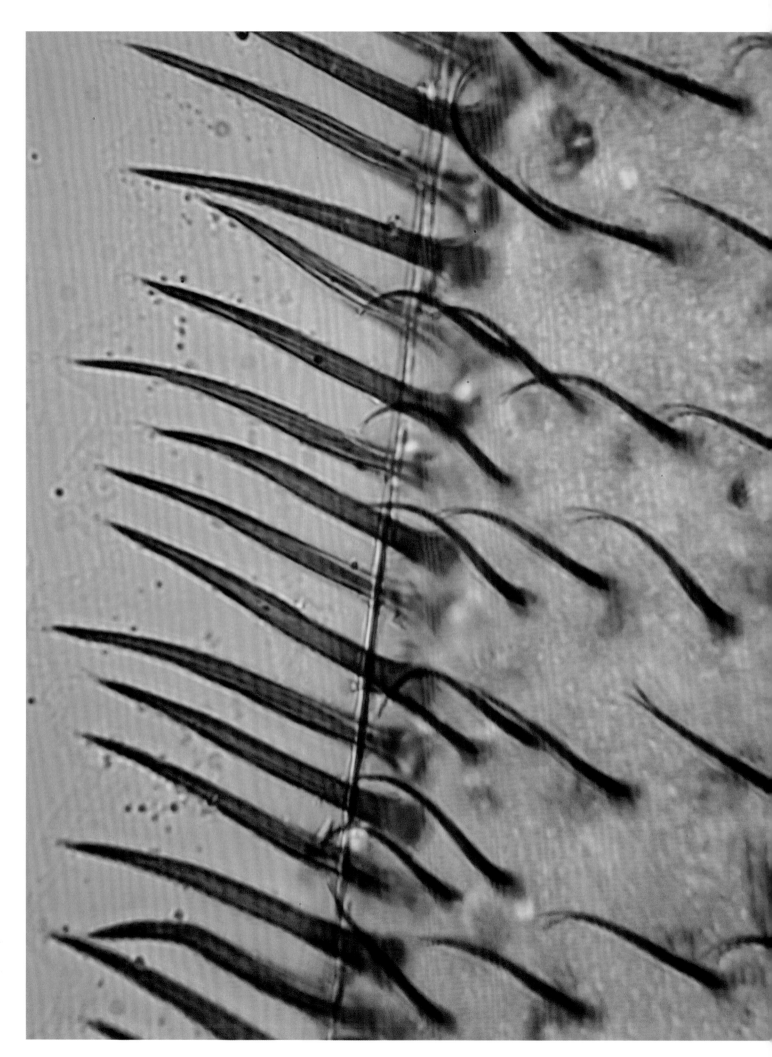

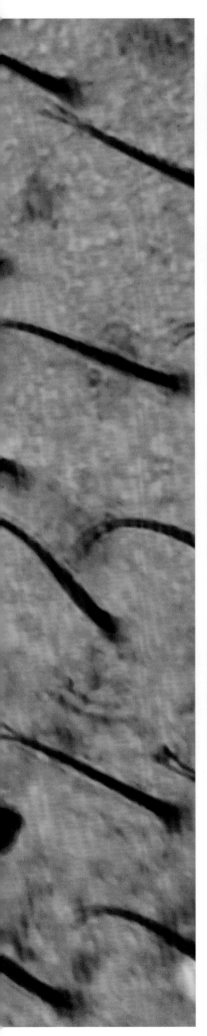

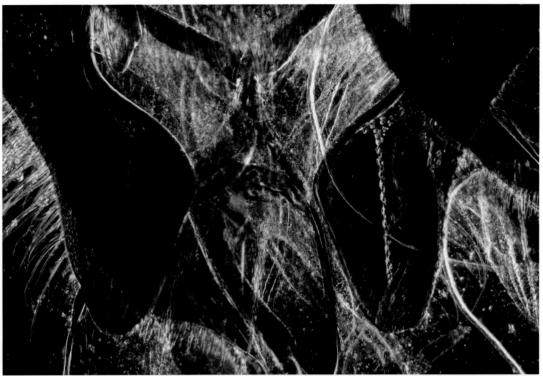

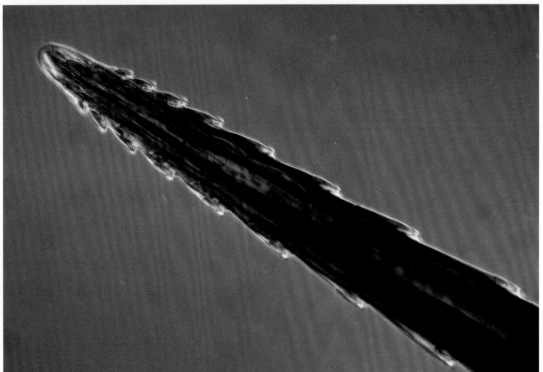

LEFT:
Waterproofing
Viewed up close, tiny hairs can be seen on a bee's wings. Quite what these hairs are for is unknown. It is possible that they work in a waterproofing capacity or they may be motion and air-flow sensors.

ABOVE TOP:
Mandibles
This micrograph, or photograph taken with a microscope, shows the pincers of the two mandibles around the top of the mouth.

ABOVE BOTTOM:
Saw system
Up close, the honeybee stinger looks suitably terrifying. The saw-shaped blade jabs easily into an attacker's skin and injects poison. This stimulates pain within seconds. The barbed stinger gets stuck in the flesh, increasing the irritation, but that spells certain death for the bee, who is disembowelled in the process.

RIGHT:
Perfect pollinators in peril
The bee is a pollen- and nectar-collection machine that evolved about 100 million years ago – around the same time that flowering plants first appeared. Today, the world's bees are weighed down by the unwitting impact of human behaviours that destroy the bees' habitats and poison them with pollution.

OPPOSITE:
Simple eyes
All bees have three simple eyes, or ocelli, on the top of the head behind the antenna. These organs pick up changes in light level.

In the Hive

As the popular phrase says, a hive is always a place full of activity. How could it not be when there are hundreds, probably thousands, of individual bees all working toward the same goal. Among the seeming chaos, there is complete order. The queen controls every aspect of her workers' lives using a cocktail of chemical signals. These suppress the development of the workers' sexual organs, rendering them incapable of producing viable young. If the queen dies, workers rush to divert the development of a few larvae from worker to new queen. If they fail, then the colony is doomed. Some workers can lay eggs, but in almost all species of social bee, these eggs will only develop into male drones, who do not contribute to the running of the hive. The team works hard to prevent this disaster, and the queen is seldom exposed to danger.

And that same teamwork sees the hive through the very hardest of times, conditions that would not be survivable for insects living alone. The workers are able to refine a high-calorie food – namely honey – from the sparsest of ingredients: a weak sugar water and pollen dust. Most solitary bees will die off before the winter arrives, leaving only eggs behind. However, the honey diet sustains the colony through the winter, when other food sources are too scarce for such a number of insects. And when the good times roll, the hive will grow so large that the queen's chemical control begins to wane. When this happens, the bees prepare for a swarm. Watch out.

OPPOSITE:
Having a ball
A swarm of dwarf honeybees gathers in a ball of bodies on a lime tree. The queen is protected deep inside, as the colony searches for a place to construct a new hive.

RIGHT:
Queen control
The queen bee wields a chemical control over her workers. The pheromone production goes hand in hand with the queen's fertility. Her hugely enlarged ovaries make her considerably larger than her infertile workers.

OVERLEAF:
Broodcomb
The open brood cells contain the chubby grub-like larvae that are growing inside. These young bees are about nine days old and will soon have their cells capped – as has already happened next door. Then they will pupate for two weeks and metamorphose into an adult worker.

Pupation
Row upon row of pupae are lined up within the broodcomb. Inside the pupal skin, or cocoon, the insect has become dormant as its body is undergoing a wholesale transformation.

RIGHT TOP:
Rare beast
The queen uses a store of sperm collected from several males on her nuptial flight to fertilize most of the eggs she lays. These will be females, mostly workers of course, but a few new queens are produced from time to time. Some eggs are not fertilized and develop into male drones, which are larger than workers with obviously bigger eyes.

RIGHT MIDDLE:
Nursery team
The job of looking after the broodcomb where eggs are laid and tending to the larvae as they grow belongs to the youngest worker bees. For the first two days on the job after emerging from their own cell, young workers clean the brood cells, making them ready for the next batch. Next they become nurse bees, who have the job of feeding the larvae as they grow in the cells. By the age of 12 days nurse bees have outgrown this task and move on to new roles.

RIGHT BOTTOM:
Laying time
A queen bee honeybee, seen here surrounded by nursing staff, can lay up to 2,000 eggs in a day. The eggs will hatch in three days, whereupon the nurse bees will give it a meal of royal jelly, a blend of proteins, fats and sugar, that is secreted from a gland in the worker's throat. All larvae get a few days of royal jelly, but only a larval queen is fed this concoction throughout her development.

OPPOSITE:
Exploitation
Many animals are fond of eating honey. Perhaps the most famous are the bears and honey badgers, the latter of which is sometimes led to the nest by the honeyguide bird. This recruits the bigger animal to smash open the nest so it can eat the grubs inside (not the honey). Humans have also been recruited by the honeyguide for the same purpose in the past, but for the last few thousand years, we have gone one better and kept our honeybee colonies in hives that we control, using an intimate knowledge of bee behaviours.

PREVIOUS PAGES:
Swarm
Bees swarm over an apple tree. A colony of bees is a superorganism, or a collection of organisms that works as one living thing. A swarm is a division of the superorganism, so one large colony becomes two or more new colonies living in new locations.

ABOVE:
Foragers
A white-tailed bumblebee collects food for the colony. The forager caste of social bees tend to be the older ones, who are unable to tend to larvae effectively. Bumblebee foragers are among the fastest insects around. They fly at speeds of 54km/h (33mph).

RIGHT:
Liquid fuel
Honeybees become foragers at the age of about three weeks. They slurp up nectar from the nectarine reservoirs at the base of a flower and store it in the 'honey stomach', a pouch at the top end of the gut. Back at the hive, this nectar is regurgitated into honeycomb cells.

RIGHT:
Apiculture
A beekeeper inspects a frame, one of several that slots into a hive. The frame creates a space for the bees to build a wax honeycomb, but can also be removed easily to harvest the honey without damaging the wax matrix so workers can clean it up and start to fill the cells with honey again.

OVERLEAF LEFT TOP:
Forest home
Honeybees are thought to have evolved in semi-desert habitats with extreme water restrictions and widely dispersed sources of low-quality foods. It is this ecology that gave rise to eusociality in other unrelated animals, such as termites and mole rats. However, the honeybee life plan has proven extremely adaptable, and hives can survive just about anywhere from forests to meadows, from the edge of the tundra to the desert.

OVERLEAF LEFT MIDDLE:
Farmers' assistants
Honeybees provide crucial pollination services to farmers growing crops, such as in this field of rape. Mobile hives can be brought onto farms to perform this task at the right time each year.

OVERLEAF LEFT BOTTOM:
Honey factory
Each one of these beehives in Ukraine could produce 25kg (55lb) of honey in one season, although the average is nearer to 11kg (24lb).

OVERLEAF RIGHT:
Queen cups
An apiarist tends to plastic queen cups containing queen bees' larvae, which are used to found new hives. In the wild, nurse bees supply these larvae with frequent doses of royal jelly, which drive their development into fully fertile females.

New home needed
These dwarf bees have just set out on their swarming adventure. Tropical bees, such as these, will swarm more often, because they are more likely to be driven to divide the colony due to high temperatures inside or a lack of water for all those workers.

Comings and goings
Forager bees must collectively fly a total of 86,000km (53,000 miles) to produce just 500g (1.1lb) of honey. This is a combined flight equivalent to 2.2 times around the world.

ABOVE TOP:
The drone
A nurse bee is dwarfed by her drone brother. He will leave the colony with his fellow brothers and a virgin queen sister. They will scatter to avoid breeding among themselves. The drone will get one chance to breed, while the queen will have several mates before returning to the colony to take over the egg-production responsibilities.

ABOVE BOTTOM:
Buzz off
The drone cannot contribute to the maintenance of the hive. Nor can he forage food by himself. As the winter approaches, there will be no new queen ready until next year, so the workers use their stings to drive the useless (and defenceless) male from the nest, to starve or die of cold outside alone.

RIGHT:
Pollen baskets
A honeybee forager has loaded up on pollen from an orange dahlia. The pollen is packed for the flight home in 'baskets' on the hind legs. These are a flattened area of the femur or 'thigh' with hooked protrusions that hold a mass of the sticky grains. The pollen basket is one possible derivation of the idiom 'bees' knees', as they represent the most important and refined part of the bee.

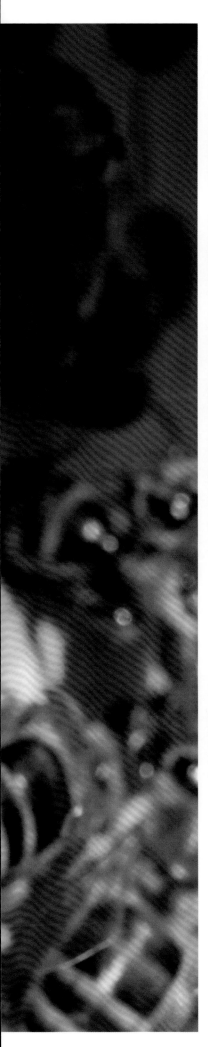

LEFT:
Food restrictions
This hive has been designed so returning forager bees can just squeeze through holes but most of the pollen is stripped from their legs in the process.

ABOVE:
Pollen trap
Supplies of pollen collected from bees build up in a pollen trap. Beekeepers will use this supply for feeding to hives early in the spring to kick-start honey production when flowers are still scarce.

OVERLEAF:
Out with the old
A honeybee swarm is normally led by the older queen. She leaves a new generation of queens developing in her old nest, and recruits half the workers under her banner to fly off and set up a new nest somewhere else.

Communal living
Solitary mason bees often live communally, especially where a kindly person has set up an ideal habitat for their mud-lined nests. Despite being crowded together, all these bees are devoting their energies to ensuring the survival of their own young, not those of another.

ABOVE:
Meeting the cousins
An orchid bee feeds on a verbena watched by a worker ant (top) – tiny by comparison. The ant lives in a eusocial system just like a honeybee (but not an orchid bee) and, despite lacking wings, is a close relative of the bees.

RIGHT:
Nectar pump
The bees tongue, or glossa, is sheathed in a pair of maxillae, creating a near tube. The glossa flits up and down inside this tube to lift nectar up to the mouth.

OVERLEAF:
Search area
Bees flying into the hive deserve a good rest – but seldom get it. The foragers will search for food as far as 6km (3.6 miles) from the hive. Any further than that and the workers expend more energy bringing in the food than is available in the pollen and nectar.

Mud hives
Old-style mud beehives seen here in Germany work perfectly well but have to be destroyed to access the honey made inside.

Honey production
Nectar and pollen collected by the foragers is received on the honeycomb and stored in cells. The nectar dosed with a little pollen and wax is then reduced into a syrupy honey. This is achieved by workers fanning the liquid with their wings, driving off the water so it steadily thickens into a substance that is so sweet that it is relatively impervious to decay, just like fruit preserves and jams.

LEFT:
Bearding
A bee swarm will frequently 'beard' before setting off. This behaviour sees the bees gather near to the old nest, forming a distinct group that then sets off on its mission to find new lodgings.

ABOVE:
Nuisance call
Bees will set up home in any suitable cavity, be it in a hollow tree branch or artificial structure. A nest near to where people live needs to be removed because constant conflict between the two home owners is inevitable.

OPPOSITE TOP LEFT:
Queen cup
The brood cell of a queen sits perpendicular to the cells of the workers. Once a colony grows so large that the queen's chemical influence loses its power, she will produce some new queens, and prepare to leave the nest for a daughter to inherit.

OPPOSITE BOTTOM LEFT:
Future rivals
Three queen pupae are close to splitting open so the adult bee can emerge. The first one out will seek out her rival sisters and sting them to death within their cocoon. There can be only one queen in the hive.

ABOVE:
New leader
The occupant of the queen cup has been fed royal jelly throughout her development, which takes a few more days than a worker. This requires a specialist team of nurse bees. Nurse bees can only produce royal jelly for nine days before their glands atrophy. Once the queen has emerged she is cleaned and fed by attendants.

OVERLEAF:
Insect hotel
This collection of hollows, nooks and crevices has been provided as a home for bees and other insects. An insect house – or a hotel on this grand scale – provides the foundation for a healthy population of insects in a habitat.

RIGHT TOP:
Not bumbling at all
Bumblebees and other social bees exhibit a behaviour called lower constancy, which means they tend to return time and again to the same blooms until there is no nectar or pollen left.

RIGHT BOTTOM:
Wide choice
Honeybees are not beholden to any flower species in particular and are able to gather food from all shapes and sizes of bloom. The bees of some hives even collect the honeydew in place of nectar. This is a sweet liquid excreted from the anuses of sap-sucking aphids.

OPPOSITE:
Sea holly
Bumblebees, such as the white-tailed species *Bombus lucorum*, have a close association with sea holly and other members of the wild carrot family. These plants often have dense clusters of small flowers that create a one-stop shop for all the bee's needs.

PREVIOUS PAGES:
Africanized bee
The Africanized honeybee is a artificial hybridization of African and European subspecies of *Apis mellifera*. This was carried out in 1956 in Brazil but several swarms of the new breed escaped into the wild. There is no significant physical difference between natural subspecies and this new strain, which has since spread across warmer areas of South and Central America and into the southwest of the United States.

RIGHT:
Danger?
The Africanized bee is sometimes called the killer bee. Between 500 and 1,000 bee stings in a short period carries a serious risk of death, and indeed people have been killed by Africanized bees. However, this is not because their stings are deadlier, but because they are more likely to attack a person nearing their nest. Deaths are extremely rare and only occur if the victim cannot escape – and people suffer the same fate when natural strains attack on occasion. So killer bees are typically not a threat to life, but resist beekeeping – the queens swarm too often – and disrupt native bee breeds as they spread into new areas.

Construction site
Wild worker honeybees are building a new home on a tree branch using yellow orange beeswax. The beeswax is secreted from glands in the abdomen. Wax workers are generally nurse bees that are too old to produce royal jelly anymore.

OPPOSITE:
Water cooler
It can be hot inside a hive, so hot it kills the brood. As the heat rises, some workers will leave the nest to make more room for cooling breezes, and other will increase airflow by fanning their wings. Foragers switch from nectar to water and slurp up a bellyful to squirt on to the honeycomb back at the hive.

LEFT TOP:
Cutting losses
A new bee nest has been abandoned after seven busy days of construction, and with the comb cells largely empty. A queen will abscond like this when there is a lack of food sources nearby. A laying queen is too heavy to fly far, so she needs to shape up prior to leaving. She stops being fed, loses weight and shuts down her egg production, ready for the journey ahead. This is a risky time for the colony.

LEFT BOTTOM:
Wild living
A wild honeybee nest will have several combs of cells with gallery space in between, a platform for the workers.

Beeswax
Workers busy themselves building cells. Each bee has eight wax glands, and these secrete a clear, colourless wax, which is a mixture of fatty acids and complex alcohols. They chew this into a more workable pulp and the yellow colour comes from the pollen grains that infect the workers and the nest.

Wasp attack
A bumblebee has met its match and is under attack from a parasitic wasp. The wasp looks like it has an immense sting but really that is an ovipositor, or tube for delivering an egg inside the body of the bee. The bee is paralysed by the attack and will become a zombie – still alive but immobile – which is slowly eaten from the inside by the wasp larvae that hatches inside.

BELOW TOP:
Sealant
A close-up view of the gap between frames in a hive show workers smearing propolis along the opening in an attempt to seal up the hole. Propolis is based on resins collected from nearby plants that are then mixed with enzymes. As well as physically filling a hole, propolis also has an anti-fungal effect.

BELOW BOTTOM:
Fire?
A beekeeper's smoke box signals to the bees that the area is on fire – they aren't to know that they live in a box. The approaching fire stimulates the bees to hurry inside and eat up as much honey as possible before it is destroyed. This imperative overrides any instinct to protect the nest and attack the invading human.

RIGHT:
Protection
Beekeepers put themselves in harm's way to collect honey. Some have learned to be gentle enough not to invite attack from their captive bees, but most prefer to wear coveralls and a net face screen to minimize the stings.

ABOVE TOP:
Ready-made home
The vertically hung framed beehive design dates from the mid-19th century. The frame makes it possible to reuse the wax honeycomb. That means that beekeepers maximize honey yields because a wax worker bee must eat 8g (0.3oz) of honey to produce 1g (0.04oz) of beeswax.

ABOVE BOTTOM:
Frames
In traditional framed hives, the frames are made from inexpensive softwood that is untreated so it does not contaminate the honey.

RIGHT:
Winter feasting
Beekeepers must leave some honey in the hive for the colony to eat during winter. It can get so cold that the bees are unable to move easily from frame to frame, so before the cold weather hits, the keepers move all the full frames together to ensure the bees have access to the food they need.

LEFT:
Stingless beekeeping
A stingless beehive is very different to that of a honeybee. A covered tray sits atop a tree stump occupied by a stingless bee nest. The bees collect their honey in pots in the tray, which are then pumped out by the beekeeper.

ABOVE:
Honey hunter
A honey hunter in the forests of Cameroon wears traditional protective clothing while harvesting wild honey from a nest in a tree trunk.

OVERLEAF:
Heave
A construction team of worker bees repairs a damaged comb with an active brood. To haul the damaged structure together, the bees must 'festoon' forming, a living sheet of bodies linked together by their tarsal claws.

LEFT:
Harvest time
The smoke box subdues the bees as the frame is prepared for the honey to be removed.

BELOW:
Uncapping
A beekeeper scrapes away the wax caps that close off the honeycomb so the honey can flow out.

BELOW:
Honey press
A simple technique for extracting the honey is crush and strain, where the whole comb is mashed up and the honey filtered out from the mixture. Here honey extracted by a spinner is sieved to filter out fragments of wax.

RIGHT:
Spin system
A less destructive method of collecting honey is to use a spinning extractor machine. Centrifugal force pushes the honey outward from the cells as the frames are spun. This oozing honey then trickles into the base for collection.

OVERLEAF:
Ancient system
Ethiopian beekeepers living near Lake Chamo have hung traditionally-built hives made from long woven baskets from an acacia tree.

Bees and Flowers

Bees form the Anthophila grouping within the wider insect order, and that name means 'flower lover'. This is a testament to how these little insects are all specialist flower feeders, visiting to feast on the nectar and pollen provided by the blooms. Many species have adapted to exploit certain kinds of plant, even going as far as collecting the oils to anoint themselves with an alluring scent. In much the same way, plants have evolved to exploit their visitors, and today thousands of species rely on bees to pollinate them, including about a third of the commercial crops grown around the world – everything from onions and potatoes to coconuts and cashews. An insect-pollinated flower produces nectar purely to lure in bees and other creatures. It serves no other purpose. The pollen is especially sticky, unlike the grains of a wind-pollinated plant, which are microscopic to aid their dispersal. The plants must tolerate the majority of their pollen being taken away by bees as food. But a few grains will make it onto the legs and body of the roving insect caller, and cling there on the journey to the next flower or the next after that, where it is brushed on to the stigma of the flower, a sticky spatula for collecting incoming pollen. Assuming the pollen is from the right plant, it will begin a journey tunnelling into the heart of the flower to fertilize the plant's ovules or eggs. This is the first step in creating seeds and fruits that will transform into next year's blooms.

OPPOSITE:
For the bees
A dwarf honeybee, *Apis florea*, collects nectar in a tropical bloom in Southeast Asia.

Hover and lick
An orchid bee, *Euglossa intersecta*, hovers near a flower in the Amazon basin using its impressively long tongue, which is far longer than the mouthparts of other bees.

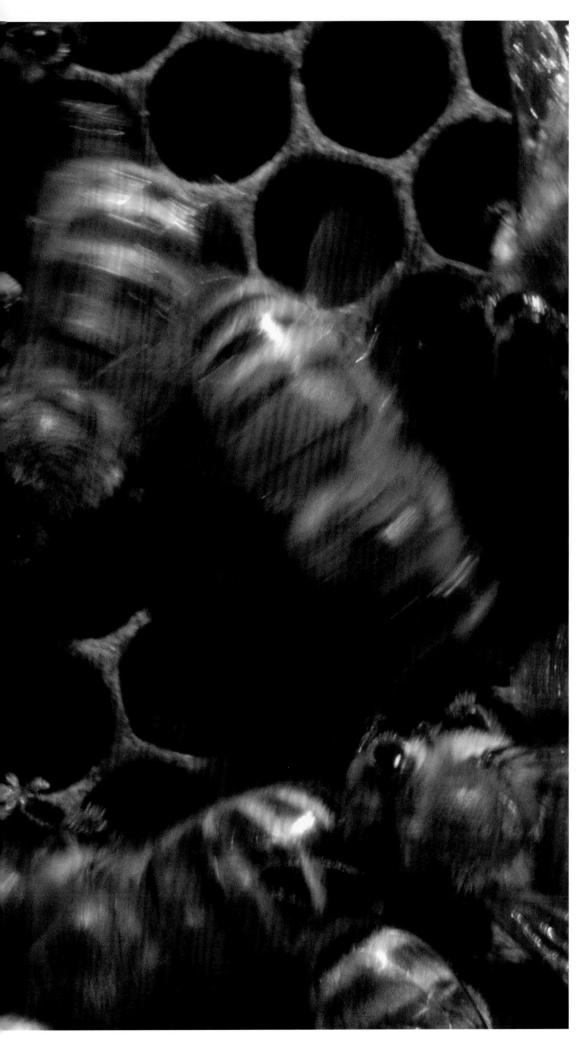

Waggle dance
A honeybee forager uses a waggle dance performed on the honeycomb to communicate the direction and distance from the hive to a good source of food. The direction is defined by the angle of the dance relative to the sun, and the number of waggles shows how far to fly. The surrounding bees pick up the information and set off to the food site.

Landing zone
This daisy offers a wide and stable landing area for the bees – more than one at a time – to gather food for the colony.

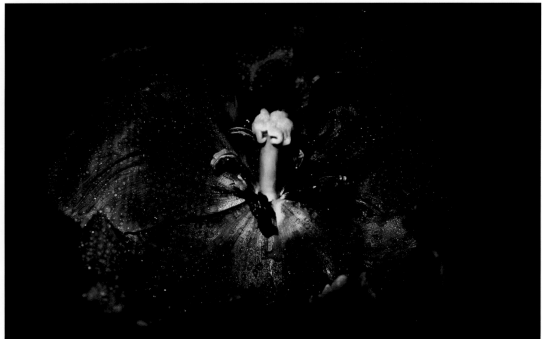

LEFT:
Slurping nectar
This orchid bee, *Euglossa imperialis*, is slurping up nectar from a tube-shaped flower in Costa Rica.

ABOVE TOP:
Bee's-eye view
This tulip flower coated in water drops shows, under ultraviolet light, more of what a bee sees as it searches for food. The pollen and nectar-rich parts of the flower stand out against the petals.

ABOVE BOTTOM:
Colour changes
In natural light this will look like a yellow flower to our eyes. But bees can detect some ultraviolet light and so to them the flower appears to have distinct zones that direct the insects to land in the middle.

Easy life
A cuckoo bee enjoys an easy life as it feeds on a coral vine. It need only feed itself, slurping up the choicest nectars and nibbling on the best pollens, unencumbered by the need to collect food for its young. Another bee species is doing that job for it!

ABOVE:
Resource management
A squad of foragers from a honeybee hive work together to strip the food resources from a head of blooms. If they do not, then a neighbouring colony will have it all for sure.

RIGHT:
Honey stomach
A western honeybee is ingesting this flower's nectar. It puts the liquid in the proventriculus, a pouch just above the stomach. In here the nectar is not digested very rapidly, and is then regurgitated into cells back at the hive.

LEFT:
Happy accident
A honeybee has a face full of pollen after making the most of a yellow flower. The transfer of pollen from flower to flower is left to chance, but if the flower attracts enough visitors the odds are narrowed.

ABOVE TOP:
Fully laden
A bee hovering near a pussy willow blossom has a good supply of pollen in its baskets. These tough foragers can carry up to 80 percent of their body weight in pollen.

ABOVE BOTTOM:
Taste sensation
The bee uses its antennae and mouthparts to taste the foodstuffs, scanning for the chemicals that might indicate fungus, mites or another infestation that could damage the hive.

Hanging food
Bees are highly adaptable and can make use of all kinds of flowers. Here they use their claws to cling to flimsy inflorescences.

ABOVE:
Orchid partnership
A metallic orchid bee searches for nectar in an orchid. Only these bees visit these esoteric blooms, which will not pollinate without their shiny green partner.

OPPOSITE:
Haulage
Once a forager, always a forager. The magnified image shows that honeybees are covered in specks of pollen that are an inevitable result of its job. Foraging is the last role of a worker. Most die at the age of seven weeks.

LEFT:
Brush past
A red hibiscus has cleverly evolved so the hefty bumblebee cannot get to the nectar without rubbing up against the flower's anthers, or pollen holders.

ABOVE TOP:
Digging deep
A honeybee wriggles up inside a tubular flower, being dusted with pollen – and passing on grains from the last flower – as it goes.

ABOVE BOTTOM:
Thistle feast
A bumblebee takes on a blooming purple thistle, which offers a big landing pad for the insects.

OVERLEAF:
Sticky dust
A honeybee is covered in pollen from a sunflower. There is little doubt it will pass at least some to a neighbouring flower.

Picture Credits

Alamy: 28 (Nature in Stock), 64/65 (Natural History Collection), 116 (Joe Dlugo), 125 bottom (Science History Images), 138/139 (Lena Ason), 166 (Philippa Starkey), 174/175 (Gustavo Mazzarollo), 176/177 (Pamela An), 204/205 (Scott Camazine), 219 (Wildlife GmbH)

Dreamstime: 10/11 (Darkop), 26 (Matee Nuserm), 42/43 (Kendal Swart), 154/155 (Piman Khrutmuang), 158 (Helen Ifill), 170/171 (Thomas Lukassek), 178/179 (Mnsanthoshkumar), 181 bottom (Elmar Gubisch), 184/185 (OneWalker), 190/191 (Marinaamelie), 194 (Olga Lukina), 218 (Oschumar)

FLPA/Biosphoto: 193 (Gilles Nicolet)

FLPA/Minden Pictures: 20 & 21 bottom (Ingo Arndt), 21 top (Albert de Wilde), 74 (Suzi Eszterhas), 75 bottom (Michael Durham), 92 bottom (Albert de Wilde), 93 (Silvia Reiche), 107 & 122/123 (Ingo Arndt), 202/203 (Konrad Wothe), 208 (Cyril Ruoso), 222/223 (Ingo Arndt)

Getty Images: 24/25 (Barcroft Media), 137 (Sameer Al-Doumy)

iStock: 89 (Dev1179), 90/91 (Matthew_Savage)

Shutterstock: 1 top (phichak), 1 bottom (Eva Mira), 5 top left (alle), 5 top right (phichak), 5 bottom (Eva Mira), 6 (Cathy Withers-Clarke), 7 (Darios), 8 (MarinaGreen), 12 (heinstirred), 13 (Enid Versfeld), 14/15 (Kevin Farr Foto), 16 (IhorM), 17 (annaj77), 18/19 (Muhammad Naaim), 22/23 (aDam Wildlife), 27 (Sacraframeart), 29 (yod67), 30 (Plamen Galabov), 31 (McCarthy's PhotoWorks), 32 (Anasteziia), 33 top & bottom (DanDKelly), 34 top (guraydere), 34 middle (manfredxy), 34 bottom (Ariel Bravy), 35 (Maciej Olszewski), 36/37 (Steven Ellingson), 38/39 (Bruce MacQueen), 40 (Jeffry haryanto kurniawan), 41 top (Jari Sokka), 41 bottom (Nilambarim), 44/45 (thatmacroguy), 46/47 (Edgar Rene Ruiz Lopez), 48/49 (DrSam), 50/51 (Kallen1979), 52/53 (Wagner Campelo), 54 top (pittawut), 54 middle (Janelle Lugge), 54 bottom (Azami Adiputera), 55 (Radin Hasanudin), 56/57 (Yusnizam Yusaf), 58 (Eatara Photographer), 59 (Niney Azman), 60/61 (pitaksin), 62 (Murilo Mazzo), 66 (Sergey), 67 top (Cornel Constantin), 67 bottom (Doni Kirana), 68/69 (Love Lego), 70 (Yuttana Joe), 71 top (Supachai sopaporn), 71 bottom (SweetCrisis), 72 top (MawardiBahar), 72 bottom, 73 (Farid Irzandi), 75 top (ermi), 76/77 (Geza Farkas), 78/79, 80/81 (marc sims), 82 (Paul Reeves Photography), 83 top (Tarahiran-on Notanat), 83 (fendercapture), 84 top (Jennifer Basvert), 84 middle (tasnenad), 84 bottom (Andreas H), 85 (CaptainJakir), 86/87 (thatmacroguy), 88 (StGrafix), 92 top (Henrik Larsson), 94/95 (Young Swee Ming), 96 top (SonyGuy), 96 bottom (Barbara Storms), 97 (d murk photographs), 98/99 (Steven Ellingson), 100 (Love Lego), 102/103 (Elvira Tursynbayeva), 104 top (Love Lego), 105 top, 104/105 bottom, 106 (Love Lego), 107 top (unol), 108/109 (Eugene B-Sov), 110, 111 top, 111 bottom (Julia_585),112/113 (ileana_bt), 114–115 all, 117 (Ileana_bt), 118/119 (mykhailo pavlenko), 120 top (Eileen Kumpf), 120 bottom (Frank Reiser), 121 (arda savasciogullari), 124 (Ileana_bt), 125 top (Pan Xunbin), 126 (Elvira Tursynbayeva), 127 (Love Lego), 128 (Mrs.Rungnapa akthaisong), 130/131 (Diyana Dimitrova), 132/133 (Vova Shevchuk), 134/135 (bamgraphy), 136 top & bottom (Ivan Marjanovic), 136 middle (rtbilder), 140 (Andrew Linscott), 141 (Maria T Hoffman), 142/143 (Roman023_photography), 144 top (Olga Samostrova), 144 middle (Angyalosi Beata), 144 bottom (Shyshko Oleksandr), 145 (Lipatova Maryna), 146/147 (tavizta), 148/149 (Zvone), 150 top (paisalphoto), 150 bottom (Irina Kozorog), 151 (sumikophoto), 152 & 153 (Ihor Hvozdetskyi), 156/157 (Dmytro Lopatenko), 159, 160/161 (Bo1982), 162/163 (Juri82), 164/165 (Dredger), 167 (Sketchart), 168 top (Jay Ondreicka), 168 bottom (Kuttelvaserova Stuchelova), 169 (Lehrer), 172 top (Vincent Mordrelle), 172 bottom (Olivier Laurent Photos), 173 (Rusana Krasteva), 180 (James Ronald), 181 top (Phatranist Kerddaeng), 182/183 (Volodymyr Burdiak), 186 top (Ihor Hvozdetskyi), 186 bottom (Mateusz Atroszko), 187 (santypan), 188 top (Kostiantyn Kravchenko), 188 bottom (Darios), 189 (ch_ch), 192 (Muhd Imran Ismail), 195 (nata-lunata), 196 (Aleksandr Gavrilychev), 197 (J. Lekavicius), 198/199 (Artush), 200 (yod67), 206/207 (szefei), 209 top (Druzchenko Olga), 209 bottom (Tanya Dimitrova Ivanova), 210/211, 212 (Darknessss), 213 (Daniel Prudek), 214 (Kirill Demchenko), 215 top (Dave Massey), 215 bottom (Mr.Background), 216/217 (Ruksatakarn studio), 220 (gertvansanten), 221 top (Almarina), 221 bottom (Simona Chira)